This book was published thanks to free support and training from:

Skunk Works Publishing School

Cultivating Blueberries:

A guide to growing blueberries in pots

J.C. FLORES

WHY I WROTE THIS BOOK

I wrote this book because...

I wanted to give people learning how to grow Blueberries the best chances for success. I have done all the failures for you so you don't have to follow in my steps. Reading this book will give you the best chances for growing Blueberries in pots.

WHY YOU SHOULD READ THIS BOOK

This book will help you...

Understand the best solution for growing Blueberries in pots or containers. It will give you the know-how on how to be successful in picking your Blueberry Variety. The soil that you will need to have for optimal growth. The tools needed to check your P.H. in the soil, along with chilled hours for your area and planting zone. Everything from A to Z will be explained in this book.

ACKNOWLEDGMENT

TABLE OF CONTENTS

CHAPTER 1: GETTING STARTED

The first step is to understand where you live and are going to plant your blueberries. Every state has a planting zone that is listed by the temperature that your state incurred. You need to check out the USDA Plant Hardiness Zone Map; this is the United State Department of Agriculture. All you have to do is find the state where you live in. It will tell you what your planting zone is, and with this information, you can choose the correct blueberries with the proper chill hours.

BLUEBERRY VARIETY

Blueberries, in most cases, can be categorized into about 5 different types of categories:

1: Northern Highbush -You will find these in northern regions because of their hardy to cold temperatures. At full-grown length, they can reach a height of 6 to 7 feet.

2: Southern Highbush – The southern Highbush You will find these in southern regions because of their hardy to the warm temperatures; at full-grown length, they can reach the height of 6 feet.

3: Lowbush- Are somewhat hardy to the cold weather of Maine and Eastern Canada. They are small in size, sometimes called wild blueberries, with a growth size of 1 foot or 12 inches tall.

4: Rabbiteye – Need to be pollinated and are native to the southeastern parts of the U.S. They can reach a height of 9 feet at maturity.

5: Half-High: - This is a hybrid; these blueberries are made from two types of blueberries. One being a Northern Highbush and the other from a Lowbush blueberry. This is done so the berry can tolerate low temperatures of -45 to -35 degrees Fahrenheit or -43 to -37 degrees Celsius. They can reach a height of 9 feet at maturity.

CHILLED HOURS

There are many new blueberries being developed, so these are my favorite blueberries for the best chilled hours.

Southern Highbush Blueberries:

- Sunshine Blue — at 150 to 200 chill hours
- Sharpblue — at 150 to 200 chill hours
- Star — at 150 to 200 chill hours
- Sweetcrisp — at 150 to 200 chill hours
- Jewel — at 150 to 200 chill hours
- Southmoon — at 150 to 200 chill hours
- Legacy — at 200 to 300 chill hours
- Gulf Coast — at 200 to 300 chill hours
- Ozarkblue — at 200 to 300 chill hours
- Misty — at 300 to 400 chill hours

Northern Highbush Blueberries:

- Bluecrop at 800 to 1000 chill hours
- Patriot at 800 to 1000 chill hours
- Jersey at 800 to 1000 chill hours
- Duke at 800 to 1000 chill hours
- Blueray at 800 to 1000 chill hours
- Nelson at 800 to 1000 chill hours
- Hardyblue at 800 to 1000 chill hours
- Friendship at 800 to 1000 chill hours
- Bluegold at 200 to 300 chill hours
- Berkeley at 300 to 400 chill hours

Rabbiteye Blueberries:

- Briteblue at 350 to 450 chill hours
- Woodard at 350 to 450 chill hours
- Austin at 350 to 450 chill hours
- Powderblue at 350 to 450 chill hours
- Climax at 400 to 500 chill hours
- Tifblue at 400 to 500 chill hours
- Baldwin at 400 to 500 chill hours
- Delite at 400 to 500 chill hours
- Brightwell at 400 to 500 chill hours

- Bonita at 400 to 500 chill hours

Half-High Blueberries:

- Northblue at 500 to 600 chill hours
- Northcountry at 500 to 600 chill hours
- Chippewa at 500 to 600 chill hours
- Polaris at 500 to 600 chill hours
- St. Cloud at 500 to 600 chill hours

Lowbush Blueberries:

- Top Hat at 100 to 200 chill hours
- Northland at 100 to 200 chill hours
- Polaris at 100 to 200 chill hours
- Northsky at 100 to 200 chill hours
- Chandler at 100 to 200 chill hours

CHAPTER 2: PICKING YOUR LOCATION AND YOUR POT SIZE

In real-estate, they say location, location, location is key. When you're out looking for the right spot to anchor your new blueberries. What you're looking for is about six to eight hours of adequate amount of sunlight. Now, blueberries do need full sun, but this is where most people can go wrong with too much sun. Your blueberries can get "KISS by The SUN." This will result in your leaves getting burned, also called leaf scorch disease. When the sun temperature reaches over 95 degrees to 100 degrees, you want to cover your blueberries with an umbrella or shade cloth. This will give your blueberries protection from UV light from the sun rays.

WHAT SIZE POT TO USE?

When choosing your blueberry varieties, you want to get two different varieties in the same pot. You can do Three, but no less or more than three in a 25-gallon pot. A 25-gallon pot is the right size to put three or two different varieties of blueberries with the same chill hours.

FINDING THE POT

Looking for the right size pot can come from any outlet store. Or even a box store, just look for a large enough basket to hold three blueberry bushes. Remember, if you find a good basket with a solid bottom. You will have to drill a few drainage holes so water can drain out and not rot your berries roots.

POT DRAINAGE HOLE SIZE

our pot drainage hole size matters a lot for such a simple thing as planting in a pot. There are many people who go wrong here by drilling small holes in their pots. The minimum hole size for a 25-gallon pot should be 1" in size. Anything smaller will clog up the hole and cause a drainage stoppage. This will lead the blueberry bush to drown its root system, causing root rot.

DIAMETER OF THE POT MATTERS

Make sure to get a big open-mouth pot and not a thin-mouth opening. The root system of blueberries is shallow; they can grow about 12" deep and spread out horizontally. This is why you need to have a big mouth-opening pot size.

COLOR OF YOUR POT

As you choose the color of your pot, please take note some people choose color for decoration which is fine for cosmetic viewing. But the best color would be black, and this is why. The color black pulls in heat from the sun and will keep your soil warm, therefore keeping your microorganisms healthy. The color white will do the opposite and mostly keep your soil cooler and make your microorganisms think it might be winter and not have then producing as much for the blueberries bush.

CHAPTER 3. ESSENTIAL TOOLS

Here are a few tools you should have to plant blueberries or work in the garden.

Gardening gloves: Nothing beats a good pair of gloves that will save your hands when pulling weeds and handling our next tool.

The shovel: This will be needed to dig a hole or just move some dirt soil around.

Garden Hoe: This tool is great for making rolls in the soil for planting your seeds.

Pruning Shears: When it comes to cutting thin brushes and shrubs nothing works better than a good sharp pair of pruning shears.

SPADING FORK

This tool is useful when you have to get in between your plant and the soil. Great for aerating the soil or transplanting new seedlings.

Rake: This tool is good when you need to gather leaves or spread compost around. All you have to do is turn over the rake and use the flip side of the rake.

Pruning Saw: When you need to cut large branches, this is the best tool for the job.

Wheelbarrow: When you need to haul material around your garden, make it easy with a wheelbarrow.

PH SOIL METER TOOL

This is the most important tool that you will need in your garden toolbox. The reason why is that blueberries need acidity. If blueberries don't have acidic soil at a range of 4.5 to 5.5 the blueberries will not produce berries. The leaves will turn yellow and will die. Not making it to the next harvest season. All your hard work will be lost, and the thought of not having blueberries will be heartfelt.

WHICH PH TOOL TO GET

There are many different types of tools to choose from for getting you the most accurate measurement of PH in your soil. Like the….

Test Strips: This shows you the color that you have to match. These are not that accurate but will get you in the range for the cost of the strip. These work well for checking water PH, but we are checking soil PH.

Soil Meter: They are also known as a 3-in-1. They check for wetness or moistness in the soil, as well as light and PH. These are found in your box store priced around $9.99. The accuracy is very poor and, in most cases, will give you a false reading.

PH METER DIGITAL PH TESTER:

You need to be careful with a tool like this because this tool is meant to check PH in water not in soil.

CAREFUL WHAT TOOL YOU BUY FOR PH CHECKING:

There are many tools out on the market and even on Amazon. You need to research and find out from others which would be the best tool to buy. You don't want to waste your money and time by buying a PH tool that you will not be happy with in the end. Making the wrong choice base on price will cost you more in the end. When you have to replace the tool, you're not happy with your choice purchase. Then, having to buy the more expanses tool in the end. It will make you feel bad about trying to save a few bucks, only to realize you made a mistake that cost you more when buying the correct tool.

CHAPTER 4. THE BLUE-LAB SOIL PEN

What does P.H. stand for? (Potential Hydrogen)

If you're going to grow Blueberries, this is the one tool you're going to love! This is the tool that I have been using for the last 4 years of my 6 years of growing blueberries since I started in 2018. I bought mine at Blue Lab Soil pH Pen | Blue Lab USA

I don't know what I paid for back then, but it's still a very good and accurate tool for checking PH in the soil. It will also give you the temperature of the soil. It comes with a probe to pre-make your holes so you won't break your soil pen. It's a well-fitted small hand-held instrument to measure your soil PH and soil temperature.

Effective in most growing media and in water/solution

The device is fully waterproof, not just water-resistant, and comes with an industry-leading one-year warranty for peace of mind. It features a hold reading function to capture measurements with ease and a simple two-point calibration for increased accuracy. The auto-off function helps to extend battery life, and automatic temperature compensation ensures accurate readings in any environment. Additionally, the device includes a AAA alkaline battery, allowing you to simply power on and go.

CLEANING YOUR BLUE-LAB SOIL PEN

There is a saying that the company tells its customers when you buy your Blue-Lab Soil Pen. (If It Dries It Dies) What this means is that you clean your blue-lab soil pen after you use it for testing. You must store it in the "Storing Solution."

CALIBRATING YOUR BLUE LAB PEN

Calibration is a two-step process. First, clean your pH pen and then calibrate to two pH points using blue lab pH buffer solutions (pH 4.0 and 7.0). The on-screen information will guide you through the process. Once complete, the calibration symbol will appear on your screen.

You should recalibrate whenever the calibration symbol disappears or after cleaning your pen. If you take daily measurements, you'll need to calibrate more frequently.

I have only calibrated my blue-lab soil pen three times in all the time I own the pen. As long as you keep it clean and in storing solution the battery is good. It will last you a lifetime.

SUPPLIES NEEDED

You get a sample kit of supplies with your blue lab soil pen. But you're going to need to buy more. I found it cheaper to get the solution on Amazon. Here is what you're going to need to get.

- PH 7.0 calibration solution

- PH 4.0 calibration solution
- Conductivity Probe Cleaner
- Storage Solution

These will be the only items you will need to keep buying to maintain your Bluelab Soil Pen in good working condition.

The more you use it, the more often you will need to clean and maybe calibrate the bluelab Soil Pen.

In the short time you might think this is a cost you might not be willing to make. But if you plan on growing Blueberries and at the cost of all the materials that you are going to put in to your blueberries. You would like your blueberry bush to grow to a healthy point to give you some berries. The cost is very small if you consider the loss of your blueberry bush due to poor PH soil. All the time and money lost to even think of starting over will be painful.

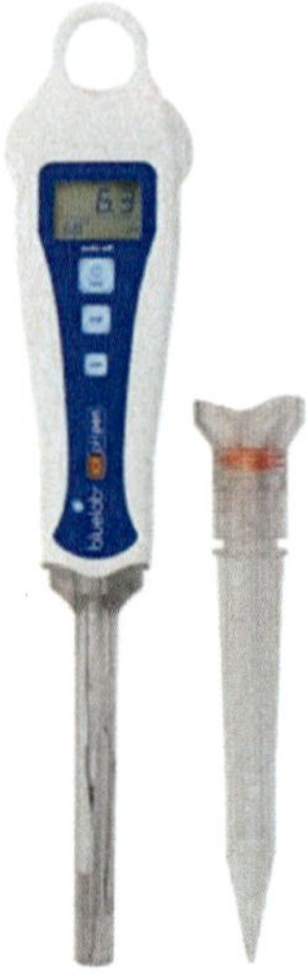

CHAPTER 5. SOIL PREPARATION

When setting up your soil, you need to understand one thing very important, which is the key to blueberries planting. This is because blueberries root system does not run deep into the soil. They are shallow root growers going down only about 12" and branch out horizontally about 6" to the side. It's for this reason that I set up my soil in a way called "Layers" or "Layering." This is the same way you would find in the wild when "Mother Nature" did it for centuries. The first layer you should start off with is **Earthgro Topsoil**. This is to bring in the worms and microorganisms, which are:

In **Topsoil**, you will find a rich and bustling ecosystem of microorganisms that thrives and gives by contribution significantly to the soil health and fertility to the soil. Let's look into the diverse cast of microorganisms that are inhabiting this rich layer.

BACTERIA

What are Bacteria?

These tiny, single-celled organisms play an essential role in giving nutrient cycling, organic matter decomposition, and disease suppression.

EXAMPLES OF BACTERIA FOUND IN SOILS

- **Pseudomonas**
- **Arthrobacter**
- **Clostridium**
- **Bacillus**
- **Micrococcus**
- **Flavobacterium**
- **Corynebacterium**
- **Sarcina**
- **Mycobacterium**

FUNGI

What are Fungi?

These filamentous organisms **break down complex organic compounds, aiding in nutrient release and soil structure formation.**

EXAMPLES OF FUNGI FOUND IN THE SOILS

- **Mycorrhizal fungi: They form a symbiotic relationship with the plant roots.**
- **Saprophytic fungi: Decompose dead organic matter.**
- **Pathogenic fungi: Some cause plant diseases, but most are beneficial.**

ACTINOMYCETES

What are Actinomycetes? These are soil-dwelling bacteria-fungi hybrids that contribute to organic matter that, when decomposed, produce antibiotics.

Examples of Actinomycetes found in Soil:

- **Streptomyces**
- **Micromonospora**
- **Nocardia**

PROTOZOA

What are Protozoa? These are single-celled eukaryotes that graze on bacteria, regulate microbial populations, and enhance nutrient cycling.

EXAMPLES OF PROTOZOA FOUND IN SOILS

- **Amoebae**
- **Ciliates**
- **Flagellates**

Blue-Green Algae (cyanobacteria):

What are Blue-Green Algae?

These are photosynthetic microorganisms that are fixed on nitrogen, stabilize soil, and contribute to organic matter.

Some Examples of Blue-Green Algae are found in Soil:

- **Anabaena**
- **Nostoc**
- **Spirulina**

VIRUSES

What are Viruses? These are non-living entities that infect other microorganisms, impacting their populations and functions.

EXAMPLES OF VIRUSES FOUND IN THE SOIL

- **Various soil-inhabiting viruses**

NEMATODES

What are Nematodes?

These are microscopic roundworms that play roles as predators, grazers, and decomposers.

EXAMPLES OF NEMATODES FOUND IN THE SOILS

- **Bacterial-feeding nematodes**
- **Fungal-feeding nematodes**

- **Predatory nematodes**

Please take note this bustling microbial community is essential for maintaining a healthy and balanced soil ecosystem, ensuring the fertility and productivity of your precious topsoil.

AFTER THE FIRST LAYER OF SOIL, ADD PEARLITE

Once you put down the first layer of topsoil, you want to add some pearlite.

- Perlite is an amorphous volcanic glass that has a relatively high water content, typically formed by the hydration of obsidian. This occurs naturally and has a way of getting bigger or expanding when heated sufficiently.

You want to add three cups full of perlite to the top of the first layer of topsoil. You don't need to mix it in; just add it to the top layer of the topsoil. This is for drainage only to have the water drain freely. This will have the blueberries roots not being wet all the time, so the plant will not drown and have the roots in a pool of water.

SECOND LAYER OF SOIL

This would be your Organic Planting Soil; there are two kinds. One for potting mix and the other for raise beds. Either one will work.

Organic Soil: This is a type of soil that forms naturally through the decomposition of plant and animal materials. Let's delve into what makes organic soil unique:

Composition: Organic soil is created by the breakdown of organic matter, such as leaves, fruits, vegetables, and animal waste. These materials enrich the soil with **nutrients** and **minerals**.

It's like the forest floor, where leaves fall, trees decay, and the animals contribute to the ecosystem. All this activity directly impacts the soil, creating a living foundation for future growth.

BENEFITS OF ORGANIC SOIL

Environmentally Friendly:

Using organic soil means returning soil made from natural ingredients back into the environment. It promotes soil sustainability and enriches your garden over time.

Healthier Plants: Organic soil helps plants resist pests and diseases. Nutrient-rich elements in the soil lead to stronger cell walls, providing added protection.

Water Retention: Adding organic material to native soil balances drainage and water retention. Organic soils keep water in the soil longer, reducing the need for frequent watering.

Chemical-Free: Organic soils eliminate the need for chemical-heavy pesticides. They naturally protect plants and make them more resistant to diseases.

VARIETIES

Organic soils come in various types, including:

Organic potting soil: Ideal for container gardening.

Lawn soil: Suitable for lawns and grass.

Garden soil: Used for planting in garden beds.

Remember, organic soil may look different from conventional or synthetic soil, but its benefits are worth it in gold. It's always best to use organic over synthetic soil. Remember, what you put in the ground, your fruits will have part of that DNA that you will eat and put into your body.

It's also suitable for trees such as magnolias, maples and conifers. Remember, this mix is ready-to-use and can be applied directly or mixed with native soil. Its organic composition ensures optimal root health and vibrant colors for your Blueberries Garden.

One big key element is just because the soil says it's acid soil. Don't rely on that information. You still need to check the PH with your Blue-Lab Pen and make sure your soil is reading 4.5 to 5.5 on the acidic scale.

Here is where many people go wrong when planting Blueberries. They rely on what the bags say and do not check for PH in the soil.

Also check your PH in the soil when planting Blueberries. This is why some people fail and others succeed in growing Blueberries.

Don't forget this is the reason why a PH meter is necessary to constantly keep checking the soil to maintain the PH at 4.5 to 5.5.

Keeping a journal and writing down your PH soil will help you achieve a good and fruitful harvest of blueberries.

In the E.B. Stone Organics Azalea, Camellia & Acid Mix:

This is a versatile blend designed specifically for acid-loving plants. Whether you're amending your soil or using it as a ready-to-use potting mix, this product caters to the unique needs of plants that thrive in acidic conditions. Here's what you'll find in this mix:

- **Aged Fir Bark:** Provides structure and aeration to the soil.
- **Aged Redwood:** Adds organic matter and improves drainage.
- **Peat Moss:** Enhances moisture retention.

- **Pumice:** A lightweight volcanic rock that aids in drainage.
- **Feather Meal:** This is a natural source of slow-release nitrogen.
- **Bat Guano:** This is rich in nutrients, especially phosphorus.
- **Kelp Meal:** This adds trace minerals and growth-promoting hormones.
- **Wetting Agent (Yucca Schidigera Saponin):** This helps with water penetration and distribution.

This mix is ideal for acid-loving plants such as azaleas, camellias, rhododendrons, ferns, begonia and blueberries. Whether you're planting directly in the ground or using a container. This is the soil mix you want to use for your blueberries.

What Happens If You Can't Find These Soil Mix

If you are in an area where they do not sell these soils. Or if shipping is just too much to pay for, then you can make your own acid soil.

By getting the organic materials I have listed in these pages you have what you need to make your own acid soil mix.

All it's going take is some time to gather all the material you need and then check for acidity with your PH meter.

By taking your time and not rushing, you will be able to achieve your goal of making your own acid soil potting mix. Just think of how happy you will be in knowing that your blueberries are very special to you when you give them to others to taste.

Creating Organic, acidic soil is essential for growing acid-loving plants like blueberries, hydrangeas, and rhododendrons. Let's explore some natural methods to achieve this:

- **Amend with Compost:** Regularly adding compost, which is rich in organic matter, gradually improves soil acidity.
- **Sphagnum Peat Moss:** Incorporate sphagnum peat moss into your soil. This will help lower your PH in the soil a little but sphagnum peat moss is only meant to help in keeping moisture in the soil.

- **Compost Tea:** Watering your plants with compost tea will organically increase soil acidity. This will also add lots of microbes that are beneficial to the soil.
- **Elemental Sulfur:** Apply organic elemental sulfur to your soil. It's a safe and efficient way to make the soil more acidic. This is a slow release, so it will take some time to reduce the PH in the soil.
- **Mulch:** Use organic mulch in your garden beds. This helps to keep moisture in the soil.
- **Coffee Grounds:** Save your coffee grounds! This can help lower a little the PH in the soil and add to the organic matter over time.

Remember, balance is crucial. While acidity benefits certain plants, excessive acid can harm or burn your plants.

Xtreme
ALL NATURAL FERTILIZER
Shrimp Meal
6-6-0
ALL NATURAL
Oyster Shell
FOXFARM
POTTING SOIL
Crab Meal
4-3-0
AZOMITE
0-0-0.2
COCO GRO
COCONUT COIR
ORGANIC
Soil Acidifier
PANDA
FILM

CHAPTER 6. CHECKING FOR 4.5 TO 5.5

Once you finish with all three layers plus the perlite. You want to make sure your soil is set and ready to receive the blueberries plants. Now, this step is very important because here is when you will find out if all the steps you have done up to this point were done correctly. This will tell you if you're going to have a good season and a nice harvest.

Watering the soil

Once your pot is complete with all three soil you want to water it not only to wet the new soil. But you want to see if you have a good drainage system in your soil. The water must drain down without pooling on the top surface in a puddle of water.

If you get a puddle of water

Oh, No! If you see that the water is not draining and a pool of water stays on the top surface of the soil. You have no choice but to start over. You have to dump the pot to expose the soil to see what went wrong. What you're looking for is clods of soil that are binning together. Maybe you did not put enough perlite in the first layer of soil, which was the Top Soil.

If You Have Drainage

Okay, good job; you have good drainage, and the water is flowing through the soil like a water park slide. Give the pot of soil a complete wet down with a good wet soak.

If You Don't Have 4.5 to 5.5 PH

So, you set up your soil, and you still don't have the correct PH. Here are some steps you can take to help you achieve the proper PH in your soil.

You can use Sulfur to help lower the PH in the soil. This is one way to bring down the high alkalinity in the soil. The sulfur comes in granules form, so it can give you a slow lease. Just sprinkle in on the top surface of the soil, and look for about one point of PH to be lower for every pound of sulfur in a 25 square foot of soil.

Don't get fooled by Aluminum Sulfate:

Aluminum Sulfate works very fast, and within days, you can lower your PH very quickly. This is because the Aluminum Sulfate goes through a chemical reaction in the soil. But the Aluminum will do harm to your plant and will make your soil toxic for the microorganisms and kill your blueberries, just for a quick fix in PH.

Some people used vinegar to bring down the PH in the soil. You can use your commend house white vinegar, which is about 5 percent acetic acid. What you want to do is take a gallon of water that has no chlorine in the water. Then, you want to add about 9-10 teaspoons to the mix. Because everyone's soil PH will be different, you need to keep good notes and write down what your soil PH is reading. Then, apply the number of teaspoons you are applying. This way, you will have a good record of how much vinegar it takes to bring your soil to blueberry scale of 4.5 to 5.5.

The last best way to bring down your Blueberries soil to 4.5 to 5.5 is with **Espoma Soil Acidifier**. This is an all-Organic way to bring down your PH. This is a granule and will give you a slow lease into the soil. Again, you will need to be aware of your PH because everyone's soil PH will be different than yours. This is why you will need to keep good notes and write down what your soil PH is reading. Then apply your number of cups of soil acidifier to write down for your records. This way you will have good notes when applying soil acidifier to the soil.

These are two of the best ways I used to bring down my PH in the soil. The reason why this is simple. I like to grow organically. So, I try very hard to stay away from chemicals because when you add chemicals to your plants. I feel you end up eating those chemicals, and that is not good for your body. Now yes, everything you buy has some form of chemical made in a lab in it. But if I can control what I grow. I can grow it safe from heavy chemicals. You have this same choice when you are growing all your fruits and vegetables.

You can take the easy way and use chemicals that were made in a laboratory to get your harvest to your table. Or you can take a slower longer way to harvest your fruits and vegetables to your table.

The choice is yours to make and there is no right or wrong way, only what you feel comfortable giving to your family and putting on the dinner table.

CHAPTER 7. PLANTING IN YOUR BLUEBERRIES

Now that you have your soil all prepared and ready to receive the blueberry plants. Go to Chapter One and order the blueberry variety that will grow in your planting zone.

PLANTING YOUR BLUEBERRY

You want to plant two to three blueberry bushes in one 25-gallon pot. Try to separate them as far apart as you can from each other. I use my hand as a gauge from the edge and try to plant them 120 degrees apart from each other. Now they will grow into each other, which is what you want. When the bees come your berries will have the best chances of getting pollinated.

STRONG ROOTS

When you dig your hole, you want to add some Mycorrhizal. You will add the mycorrhizal to the hole and to the roots of your blueberry root ball. This is an inoculant wettable power. This is a pure mycorrhizal microorganism that is alive soluble organic root enhancer for reservoirs and waterborne applications. This not only works on blueberries but is very good for vegetables, fruits, herbs and all flower plants.

NO BERRIES FOR THE FIRST YEAR

This will be the hardest thing to do to your blueberries. In the first year, you need to pluck all the blueberries from your bush. This is so the blueberry bush puts all its effort and concentration into growing strong roots and not berries for the first year. Strong Roots is what you want.

This will take another year, but just know that you will have a blueberry bush that will be very strong and will give you berries for many years for those two years of non-producing berries years.

DON'T GET KISS BY THE SUN

Remember, blueberries need full sun but nothing over 95 degrees Fahrenheit. So, you need to buy a good outdoor thermometer. You need to check the temperature every day making sure the temperature doesn't reach over 95 degrees Fahrenheit or 35 degrees Celsius.

HOW TO BLOCK THE SUN-RAYS

If the sun is too hot, you would need to give your blueberries cover shade. This can be done by buying sun cover shade to put over-head. Or you can use an umbral to give your blueberries shade. You could use anything you can find that will cut the full sun rays down to protect your blueberries from being "Kiss by The Sun."

CHAPTER 8. TAKING CARE OF BLUEBERRIES

WATERING

Now that you are done with all the steps needed to plant and grow your blueberries. You need to make sure you water your blueberries. You want to make sure your blueberries' soil is wet and not drowned in water.

YOUR PH SOIL

You need to check your PH and make sure the soil PH is in the range of 4.5 to 5.5 on the PH blueberries scale.

FEEDING YOUR BLUEBERRIES

Make sure you feed your Blueberries the minerals they need in the soil every year.

FERTILIZING THE SOIL

Fertilizing the soil with compost tea or adding organic material to the soil every year to give back to the soil what it has used up in growing the blueberries.

CHAPTER 9. PHOTOS OF MATERIALS NEEDED

NATURAL ROOT ENHANCER
Xtreme
GARDENING
MYKOS
PURE MYCORRHIZAL INOCULANT
PURE · FRESH · ALIVE
GREAT FOR VEGETABLES, FRUITS, HERBS & FLOWERS
OMRI

digz

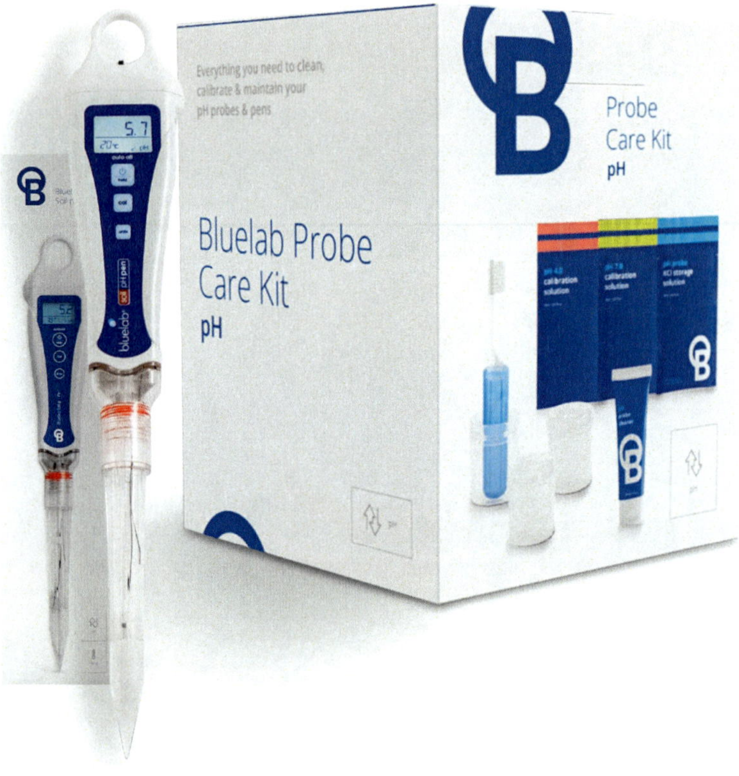
Everything you need to clean, calibrate & maintain your pH probes & pens
Bluelab Probe Care Kit
pH
Probe Care Kit
pH
5.7
bluelab
pH pen

WESTWARD

SK5

SEYMOUR S

40
20
60
0
80
-20
100
-40
-60
°F
120
TAYLOR

SPECILITE
SPECILITE
SPECILITE 5/8" Upgraded Hot Water Hose -50 °F to 190°F

CHAPTER 10. PHOTOS OF SOILS NEEDED

MUST BE PAID FOR
BEFORE LOADING.
CEDAR BARK
2 CUBIC FEET
SHREDDED
CEDAR BARK
2 CUBIC FEET
LGM

SHREDDED REDWOOD
G&P ORGANICS
Premium Topsoil
GARDEN QUALITY SOIL
CRAFTED FOR ORGANIC GARDENING
1 CUBIC FOOT (28 LITERS)
ORGANICS

G&B
ORGANICS
Purely
Compost
1.5 CUBIC FEET (42.5 LITERS)
ORGANICS
Purely Compost

ORGANICS
Harvest Supreme
ORGANICS
Harvest Supreme
MASTER
BUMPER CROP
NATURAL
ORGANIC
SOIL BUILDER
NATURAL & ORGANIC
SOIL BUILDER

Miracle-Gro
NO MIXING REQUIRED
FOR USE IN RAISED BEDS
Organic
CHOICE
RAISED BED &
IN-GROUND SOIL
WITH COMPOST
Trees, and Shrubs
Made With
Compost
OMRI
LISTED
50

Composted Chicken Manure
ORGANICS
Chicken Manure
Composted Chicken Manure
ORGANICS
ORGANICS
Composted Chicken Manure
ORGANICS
ORGANICS
ORGANICS

2-IN-1 PLANTING
MIX & MULCH
ORGANIC
& Natural
OMRI
2-IN-1 PLANTING
MIX & MULCH
2-IN-1 PLANTING
MIX & MULCH
2-IN-1 PLANTING

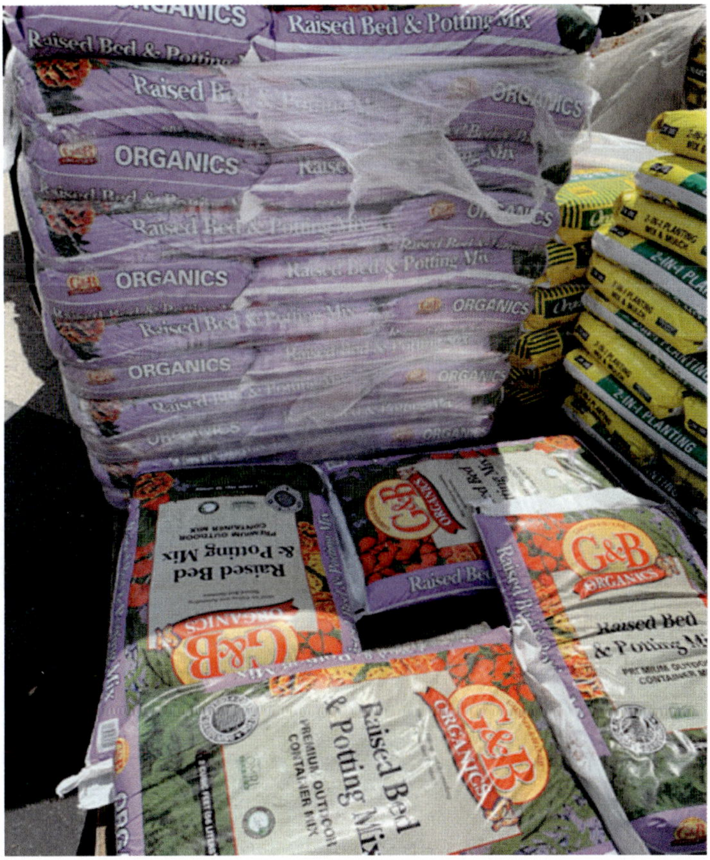
ORGANICS
Raised Bed & Potting Mix
G&B ORGANICS
Raised Bed & Potting Mix
PREMIUM OUTDOOR CONTAINER MIX
2-IN-1 PLANTING

REDI-GRO
A NATURAL FOREST PRODUCT
Walk On Bark
MULCHING:
Water penetration, moisture conservation,
soil aeration and weed control are
Coverage: One 2 cu. ft. bag covers approximately 24 sq. ft.1" deep.
INGREDIENTS: fir bark, fir chips
2 Cu. Ft.
56.6 Liters
REDI-GRO

E.B.Stone ORGANICS
AZALEA, CAMELLIA & ACID
POTTING OR PLANTING
AZALEA, CAMELLIA & ACID MIX

Potting Mix
Grows Plants Twice As Big!
Potting Mix
Potting Mix
Potting Mix
Potting Mix

EARTHWORM CASTINGS
E.B.Stone
ORGANICS

GARDENERS
SHREDDED REDWOOD
SHREDDED REDWOOD
GARDENERS
SHREDDED REDWOOD
SHREDDED REDWOOD
GARDENERS
SHREDDED REDWOOD

CHAPTER 11. ESTIMATED PRICES OF SOIL 2 CUBIC FT. OR 56 LITERS

SUCCESS
G&B
ORGANICS
Acid Planting Mix
Acid
Planting Mix
Great For Azaleas, Camellias
Rhododendrons & Other Acid
Loving Shade Plants
Acid Planting Mix
MASTER NURSERY
NATURAL
SOIL BUILDER
BUMPER
17
$11.99
QUICK COLOR
or POT PACK
Sale Price
$4.99
WE

E.B.Stone
ORGANICS
AZALEA, CAMELLIA & ACID
POTTING OR PLANTING
A UNIQUE MIX THAT IS IDEAL FOR ALL ACID-LOVING PLANTS.
$12.99
ROSE &
THE PERFECT PLANTING MIX FOR FLOWERS AND ROSE GARDENS.
Post

G&B
ORGANICS
Planting
Mix
PREMIUM GARDEN SOIL
2 CUBIC FEET (56 LITERS)
Harvest Supreme
Planting Mix
#34
10.99
2.CF

CHAPTER 12. PHOTOS OF THE BLUE

BLUEBERRIES

ABOUT THE AUTHOR

Life is good—embrace it by going outside and planting something!

J.C. Flores is a new and up and coming author, spending

his life and passion in the soil. He is driven by a passion for education and inspiration, J.C. empowers fellow authors and entrepreneurs to reach new heights and live their dream lives

Discover more about J.C. on his YouTube channel: https://www.youtube.com/@thebluegarden2060

OTHER BOOKS BY J.C. FLORES

Amazon.com: Basketball The Games Kid's Play: Without Tears In Your Eye's For Odd Number of Players eBook: Flores, J.C.: Kindle Store.

https://www.amazon.com/Basketball-Games-Kids-Play-Without-ebook/dp/B0CYYH2RTQ

https://www.amazon.com/Wildlife-Coloring-Book-J-C-Flores

Coming soon on Amazon and Ingram

How To Pass Your Written Certification, To Get Your Forklift License *Qualification Is Easy When You Learn From The Best Instructors*

ONE LAST THING...

If you enjoyed this book or found it useful, I'd be very grateful if you'd post a short review on Amazon. Your support really does make a difference and I read all the reviews personally so I can get your feedback and make this book even better.

If you'd like to leave a review, then all you need to do is click the review link on this book's page on Amazon:

Thanks again for your support!

Life is good; go outside and plant something!

Thanks again for your support!

Made in United States
Orlando, FL
04 November 2026

90200810R10069